ABCteria

An alphabetical tour through the microbial world

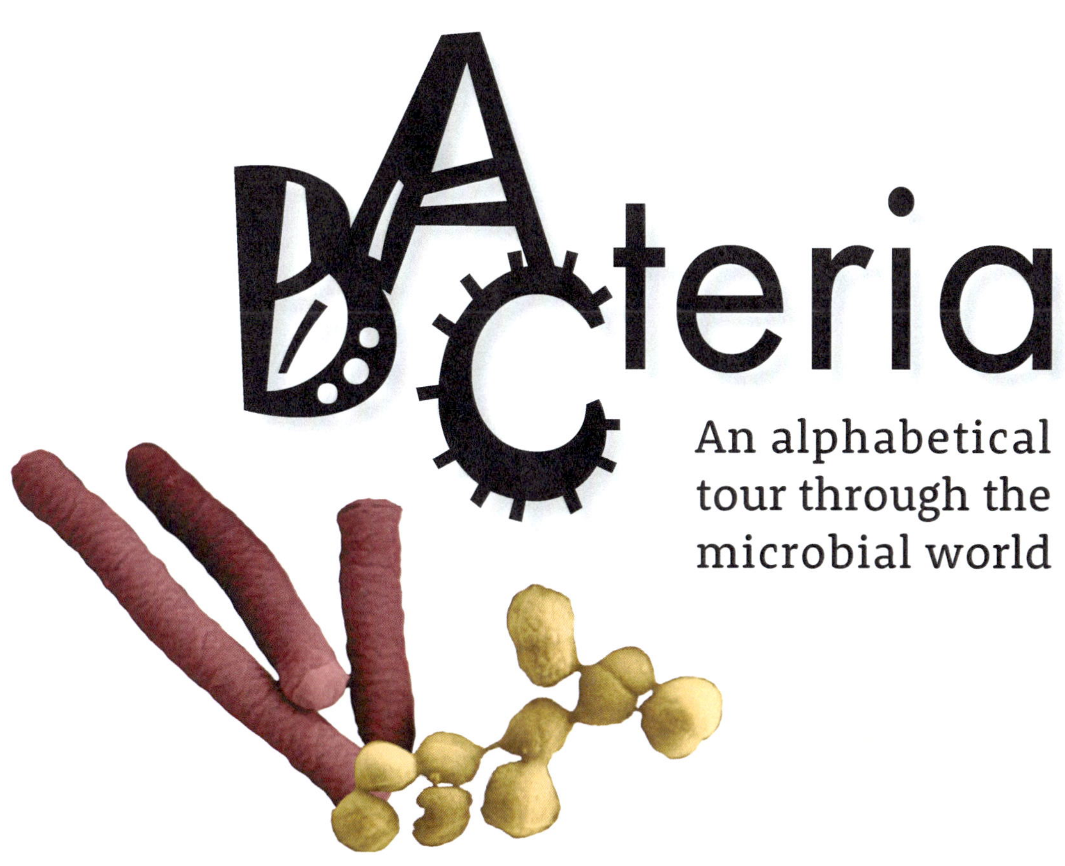

Designed and Written by

Jeanne Kagle

We live in a microbial world.

Microbes are the foundation of all life on Earth. Occupying every habitable niche on the planet, they have been evolving and diversifying for close to 4 billion years. They produce our food and nutrients, provide us with the oxygen we breathe, and degrade waste, including human-made chemicals that don't exist in nature.

In a single gram of soil there can be more than 40,000 bacterial species. The total number of bacterial species could be more than a million, possibly as many as a billion. With so many species, it shouldn't be surprising that bacteria and other microbes have a plethora of structures, survival strategies, and methods of making a living. Many seem so strange to us that they would be hard to believe in a sci-fi movie. Among all those diverse microbes, only about 2000 varieties of bacteria are known to cause disease in humans. It is this tiny minority of disease-causing bacteria, though, that get the vast majority of the public's attention.

This book is dedicated to providing an introduction to the true diversity of bacteria. As a reflection of the ecological importance of bacteria, most of the highlighted bacteria play beneficial roles in human life. In addition, several have characteristics which humans have harnessed to advance civilization and technology. The bacteria in this book are just a sample of the amazing microbial world. I hope it motivates you to learn more and appreciate the microbes all around that make our lives possible.

A

You have almost certainly witnessed the handiwork of the skilled plant engineer, *Agrobacterium*. On just a short stroll outside, you can find many plant galls, tumors, and other deformations which have a wide range of causes, from insects to injury. Many, however, are caused by *Agrobacterium*. Although it often lives in soil, *Agrobacterium* is remarkable for its ability to form tumors, specifically crown gall tumors, on a wide variety of plants. Commonly known as burls, these bacteria-induced deformations are most often found near the ground or at branch junctions.

Agrobacterium

The large, gnarly crown gall tumors we see on trees and other plants begin when microscopic *Agrobacterium* bacteria enter the plant through a wound. Once inside, the engineering feat begins. The *Agrobacterium* cells build syringe-like structures, called pili, connecting each bacterium to a plant cell. A piece of the bacterium's DNA passes through the structure and into the plant cell. Genes on the transferred DNA instruct the plant to do two things.

Some of the genes transferred to the plant cell direct it to produce plant growth hormones, causing the plant cell to reproduce uncontrollably, forming the visible tumor. Second, the plant cells containing the transferred DNA begin to make a type of chemical called opines, which *Agrobacterium* is uniquely well-adapted to consuming. By transferring DNA to the plant, *Agrobacterium* essentially turns the plant into its personal food factory.

Plant engineer

Scientists quickly recognized the opportunity to use *Agrobacterium*'s ability to transfer specific DNA to plant cells. By replacing the DNA which *Agrobacterium* naturally transfers with DNA of the scientists' choosing, they can genetically alter the plant cells as desired. Since the first attempts in the 1980s, *Agrobacterium* has become one of the most common methods of genetically engineering plants.

Bacteroides

Most of the living cells in your intestines belong to a group of bacteria called *Bacteroides*. *Bacteroides* accounts for approximately 60% of the bacteria in your intestines. The familiar *E. coli* makes up less than 1%. Because *Bacteroides* cannot live in the presence of oxygen and is finicky in its nutritional needs, it is much more difficult to grow in the lab than *E. coli* and therefore less widely studied. It is far more important, however, to maintaining human health.

Having a high proportion of *Bacteroides* in your gut bacteria is important for your overall health. Although it is easy to understand why the correct bacteria in your intestines would be important for digestive health, the types and proportions of bacteria and other microbes present in the gut have been linked to health effects in almost every body system. Changes in your gut bacteria can impact everything from blood glucose and cardiovascular health to development of the immune system and mood.

Bacteroides itself provides several health benefits. Its presence alone holds other bacteria, such as *E. coli* and its relatives, in check. *E. coli* and its relatives can lead to inflammation and prevent beneficial bacteria from properly functioning. *Bacteroides* also colonizes intestinal mucous, consuming some of the carbohydrates present. The waste products from *Bacteroides'* metabolism interact with the intestinal lining, triggering the release of hormones. These hormones decrease blood glucose and triglycerides, reduce inflammation, and increase the body's production of serotonin. *Bacteroides* is also the main source of some vitamins, particularly vitamin K, for the human body.

Consuming foods high in complex carbohydrates, such as whole grains, can encourage the growth of beneficial bacteria like *Bacteroides*.

clostridium

Without Clostridia, we would be buried in waste. Clostridia are critical to the breakdown of organic matter in both natural ecosystems and human-constructed environments. They play a key role in the decay of dead organisms in the soil. Proper decomposition of waste in compost and sewage also relies on Clostridia, especially where oxygen is not available.

Compost and Cosmetics

The endospores formed by Clostridia can survive more than half an hour in boiling water and are barely impacted by alcohol hand sanitizers. Endospores have been reliably documented as remaining viable for hundreds of years, with some accounts of endospores millions of years old germinating and being grown in culture. The main public health concern involving endospores is the spread of *C. difficile* in health care facilities. This dangerous gastrointestinal pathogen is incredibly common in hospitals and nursing homes. Typical disinfection methods don't work against *C. difficile*. Washing with soap and water is best for removing endospores from hands and bleach is necessary to kill endospores on surfaces.

There is a wide variety of Clostridium species, but they all have a few characteristics in common. Clostridia are strict anaerobes, meaning they cannot live in the presence of oxygen, and they are very good at degrading organic matter. They are also one of a few groups of bacteria that can form protective structures called endospores, which remain intact under many forms of stress. Although most play beneficial ecological roles, some make toxins which make them serious health threats, such as *C. tetani* (tetanus), *C. perfringens* (gas gangrene), *C. difficile* (commonly called C diff), and *C. botulinum* (botulism).

The cosmetic treatment Botox is actually a diluted formulation of the toxin produced by *C. botulinum*. This toxin, botulinum, is a potent neurotoxin which causes the disease botulism. It acts by binding to the nerves which trigger muscle contractions and prevents nerve signals from reaching the muscles. In botulism, it becomes impossible to contract muscles and, if left untreated, the victim eventually cannot breathe. Botox, when used cosmetically, relaxes facial muscles, reducing the appearance of wrinkles.

Deinococcus

Radiation Resister

Deinococcus could easily survive a nuclear blast. In the 1950s, a can of radiation-sterilized meat somehow spoiled. Nothing should have survived the level of radiation used. Deinococcus, however, not only survived, but thrived.

The ability of *Deinococcus radiodurans* to survive extremely high levels of radiation perplexed scientists for decades. *Deinococcus* can survive 5,000 Grays of radiation exposure, one thousand times the amount deadly to a human. For comparison, a mile from Ground Zero of the atomic bomb dropped on Hiroshima in 1945, human victims absorbed about 10 Grays of radiation.

Deinococcus is able to survive such high levels of radiation because of its extensive DNA repair system. Ionizing radiation, like that produced by an atomic bomb, is deadly because it breaks the bonds in the backbone of DNA. For most organisms, more than a couple breaks in the DNA backbone are too many to repair and the affected cells die. *Deinococcus* cells, however, each carry up to 10 identical copies of their DNA. They can use the multiple copies as templates to repair as many as 200 breaks in their DNA and essentially stitch a complete genome back together from the remaining fragments in a of couple hours.

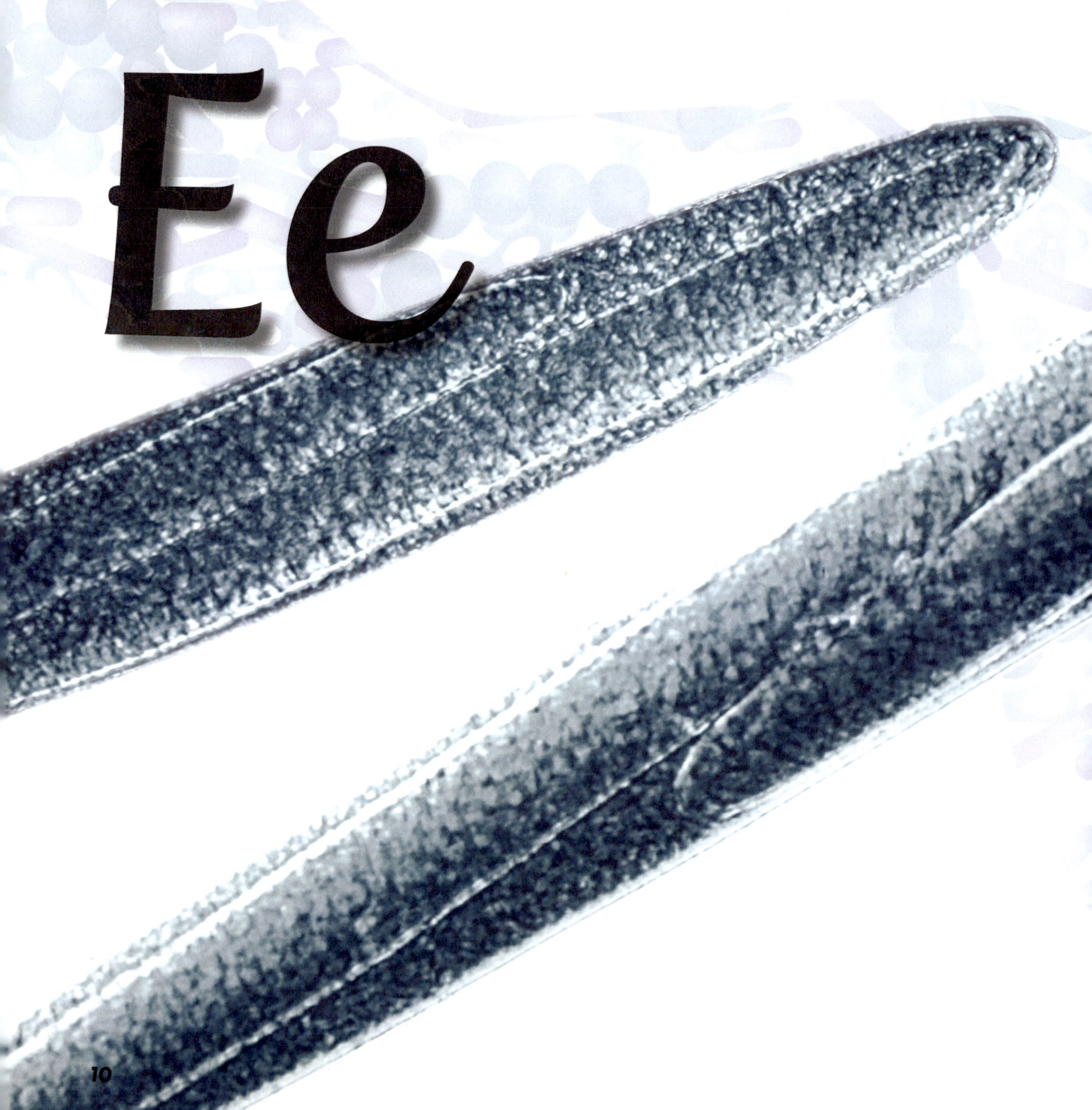
Ee

Epulopiscium

Epulopiscium fishelsoni is one of the largest bacteria ever discovered. The largest cells of this bacterium are about the size of the period at the end of this sentence.

Because of its size, when it was first discovered in the gut of tropical reef surgeonfish it was believed to be a protist. It was not until DNA analyses were available that it was definitively shown to be a bacterium. Bacteria rely on their small size for necessary tranport of materials within, into, and out of their cells. With a volume approximately 2000 times that of a typical bacterial cell, *Epulopiscium* faces a significant challenge. One way *Epulopiscium* is able to maintain its unusually large size is to have as many as 30,000 copies of its genome arrayed around the cell. This way each part of the cell has easy access to the DNA and its necessary products. *Epulopiscium*'s cell membrane is also extensively folded, increasing the surface area to allow for efficient diffusion of nutrients in and waste out.

The life cycle of *Epulopiscium* is unique among known bacteria. Because it relies on its host fish for food, its reproduction is tightly linked to the fish's circadian rhythm. In the morning, the bacterium is active along with the feeding fish. As the day progresses, endospore-like structures begin to form at each end of the cell (*Epulopiscium* is related to *Clostridium* (p. 6)). By the end of the day, these two structures grow to completely fill the original cell. Overnight, the two new cells mature and, at dawn, emerge to restart the cycle.

Guest at the banquet of a fish

FRANCISELLA

Francisella is the most infectious bacterial pathogen known.

A single cell of *Francisella* can cause serious disease.

Francisella is most frequently found in rodents and rabbits, giving the disease it causes the common name "rabbit fever". Natural *Francisella* infections are rare. Of more concern is its potential as a bioweapon. Because of its low infectious dose, it is estimated that the intentional aerosolization of 50 kg of *Francisella* in a metropolitan area could result in approximately 250,000 incapacitating infections and 20,000 deaths.

Infection can be caused by several routes, including inhalation, ingestion, tick or insect bites, and handling sick or dead animals. Depending on the route of exposure, the resulting disease can have a wide range of symptoms from minor symptoms such as sore throat and mouth sores to severe joint pain, fever, and potentially deadly pneumonia. Without antibiotic treatment, pneumonia caused by *Francisella* is fatal in about half of cases.

G
Geobacter
S

Geobacter can generate an electrical current that humans can harness. Many organisms capture energy using the transfer of electrons from the broken chemical bonds of the compounds they degrade to a different molecule. For example, in humans that molecule is oxygen. Usually, this chemical reaction happens within the cell. Bacteria like *Geobacter*, however, can transfer the electrons to an external material, like a metal electrode, creating an electric current.

Bacterial Battery

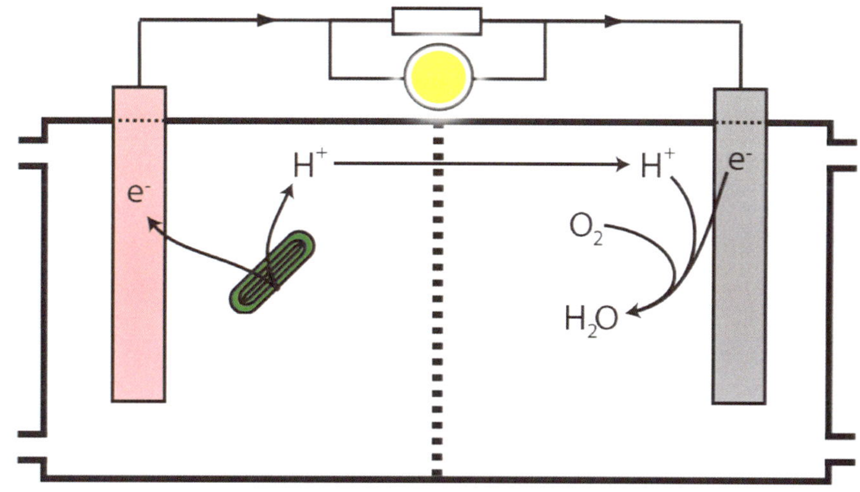

Geobacter physically attaches itself to a surface with a structure called a pilus, which acts like a microscopic wire. As part of its energy-generating metabolism, it disposes of waste electrons by passing them through the pilus to the surface. The substances fed to the bacteria and the final electron-receiving surface can be manipulated to alter the amount of voltage produced. Systems using the current-generating ability of *Geobacter* and similar bacteria are called microbial fuel cells.

Simple microbial fuel cells can be constructed from mud, wire, and an appropriate material for the bacteria to transfer electrons to, such as carbon fiber mesh. These simple fuel cells do not produce much power, though, and microbial fuel cells are currently too cumbersome to be useful to power homes or cars. Considering the need to move to methods of producing renewable energy, however, optimizing the productivity of and developing new applications for microbial fuel cells are areas of active research. One interesting approach is to use sewage or industrial waste as the input for microbial fuel cells, thereby reducing pollution and generating electricity simultaneously.

Most prokaryotes (organisms without a cell nucleus) have spherical (cocci) or rod-shaped cells (bacilli). They never have right angles...or that was the common wisdom until *Haloquadratum* was discovered in 1980. Several other prokaryotes with square cells have been discovered since, and, interestingly, they are all found in high salt environments.

Haloquadratum is not a bacterium but rather an archaeon, an evolutionarily distinct type of prokaryote. Current biological classification divides living organisms into three Domains: Bacteria, Archaea, and Eukaryotes. Bacteria and Archaea are both prokaryotes, single-celled organisms without a nucleus or membrane-bound organelles, but they are not very closely related. Archaea and Eukaryotes (the Domain to which plants, animals, and fungi belong) are actually much more closely related.

HALOQUADRATUM

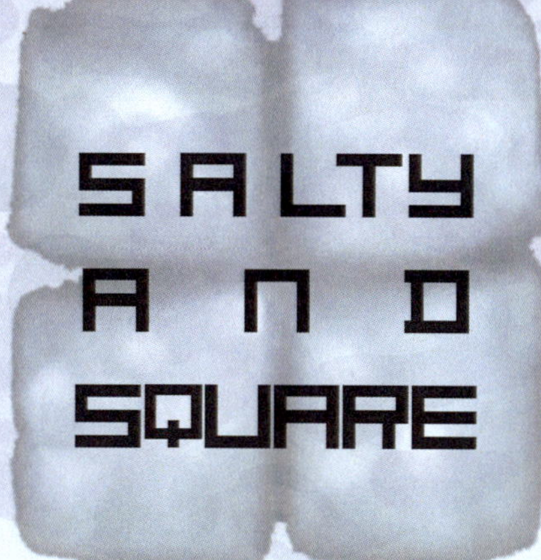

Haloquadratum and similar salt-loving archaea, have a red-purple protein in their membranes that allows them to capture light energy and convert it to a form usable for metabolic reactions in the cell. This protein, bacteriorhodopsin, is a close cousin to rhodopsin, the protein in the human eye which changes light to an electrical signal sent to the brain. Salt flats and salt evaporation ponds often look pink, red, or purple because of the archaea present and the resulting high concentration of bacteriorhodopsin.

INNUMERABLE

There are approximately 5,000,000,000,000,000,000,000,000,000,000 bacteria on Earth.

In the human body alone, there are 10 to 100 trillion bacteria. That is at least equal to, possibly many times more than, the number of human cells in the body.

MORE THAN THE STARS IN THE SKY...

For every star in the known universe, there are a million bacteria on Earth.
If you lined up all the bacteria on Earth end-to-end, they would stretch
across the Milky Way...a thousand times!

...OR GRAINS OF SAND ON THE EARTH

The number of grains sand on Earth is estimated to be 7.5×10^{18}
(7.5 billion billion). There are a trillion times as many bacteria.

Janthinobacterium is a cold-loving bacterium that has been found in a wide range of habitats from Antarctic ice to salamanders. When grown in culture, it appears deep purple due to a compound called violacein. In addition to being a beautiful color, violacein has antimicrobial properties.

Janthinobacterium

Each year millions of people suffer from bacterial infections which cannot be treated with antibiotics and tens of thousands die as a result. Violacein and other compounds produced by *Janthinobacterium* have been shown to kill or inhibit antibiotic-resistant bacteria cultured from active infections. Research into potential new treatments like violacein is critical to combat increasing bacterial resistance to existing antibiotics.

The distinctive color of violacein makes its use as a natural fabric dye an attractive possibility. When used as a dye, violacein keeps its antimicrobial effectiveness. The resulting antimicrobial textile has the rich purple color of violacein and can be used in any application where antimicrobial properties are desired.

Like all amphibians, the red-backed salamander, common in the northeastern United States, is threatened by the spread of the lethal chytrid fungus. *Janthinobacterium*, however, is often found on their skin. Salamanders harboring large numbers of the bacterium are protected from chytrid infection by the violacein and other antimicrobials it produces.

K

The microbial community of the skin is fully established just a few weeks after birth. The bacterial population shifts substantially at puberty (frequently evidenced by the development of acne) but remains surprisingly stable for the remainder of adulthood.

With you through it all

Kocuria and its close cousin *Micrococcus* are some of the most prevalent residents of human skin. They seldom cause disease in humans. On the contrary, resident bacteria create a protective barrier against pathogens and other undesirable transient flora.

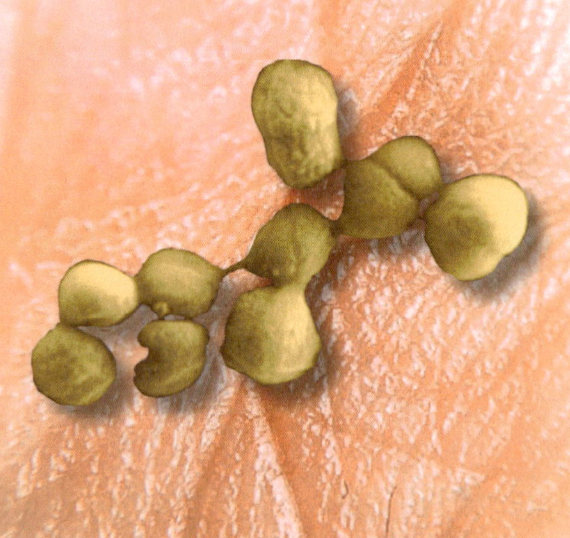

Kocuria

People shed hundreds of bacteria-laden skin cells every hour. When indoor air is sampled, about 4% of the bacteria cultured are *Micrococcus* or *Kocuria*. Because every person has their own unique skin flora, bacterial signatures are being explored as a possible forensic tool.

In addition to acne, imbalances in the skin bacterial population are associated with a variety of common skin afflictions. Skin flora is implicated in dermatitis, psoriasis, and the formation of dandruff.

Bathing and hand washing are important for removing potentially problematic transient flora, but you cannot wash off your resident flora. Soon after washing, the normal skin flora returns to its original numbers and resumes protecting you from infection and helping to maintain a healthy skin environment.

In addition to creating delicious food, fermentation is an age-old method of food preservation.

You may not realize it, but you've almost certainly eaten *Lactobacillus*. Most fermented foods (excluding alcohol and bread) use *Lactobacillus* to at least some extent. Yogurt, cheese, and sauerkraut rely heavily on *Lactobacillus*. Even making some sausages and beers involve *Lactobacillus* fermentation. *Lactobacillus* is part of a group of bacteria called lactic acid bacteria. When lactic acid bacteria extract energy from their food through fermentation, one of the main waste products is, unsurprisingly, lactic acid. This acid, along with other products made in smaller quantities, give fermented foods their characteristic sour or tart taste.

The process of fermentation also increases the nutritional value of food. The bacteria produce vitamins that would not otherwise be present. Some compounds made by fermenting bacteria, including lactic acid, have specific health benefits, such as reducing inflammation and regulating metabolism.

Because of the lactic acid that *Lactobacillus* produces during fermentation, the food becomes uninhabitable for most bacteria and fungi that cause food spoilage or disease. Some species of *Lactobacillus* also release toxins which kill bacteria that might be dangerous, increasing the safety of fermented foods.

Lactobacillus is also one of the types of bacteria found at higher numbers in healthier populations of gut microbes. Various species of *Lactobacillus* have been shown to improve digestive health, the immune response, and even mood. *Lactobacillus* is also one of the most common types of bacteria found in probiotic supplements. Although probiotics can be helpful in restoring or maintaining healthy gut flora in some situations, eating fermented foods and a diet high in whole grains is more effective for establishing healthy gut microbiota over the long term.

Lactobacillus

Fabulous Fermenter

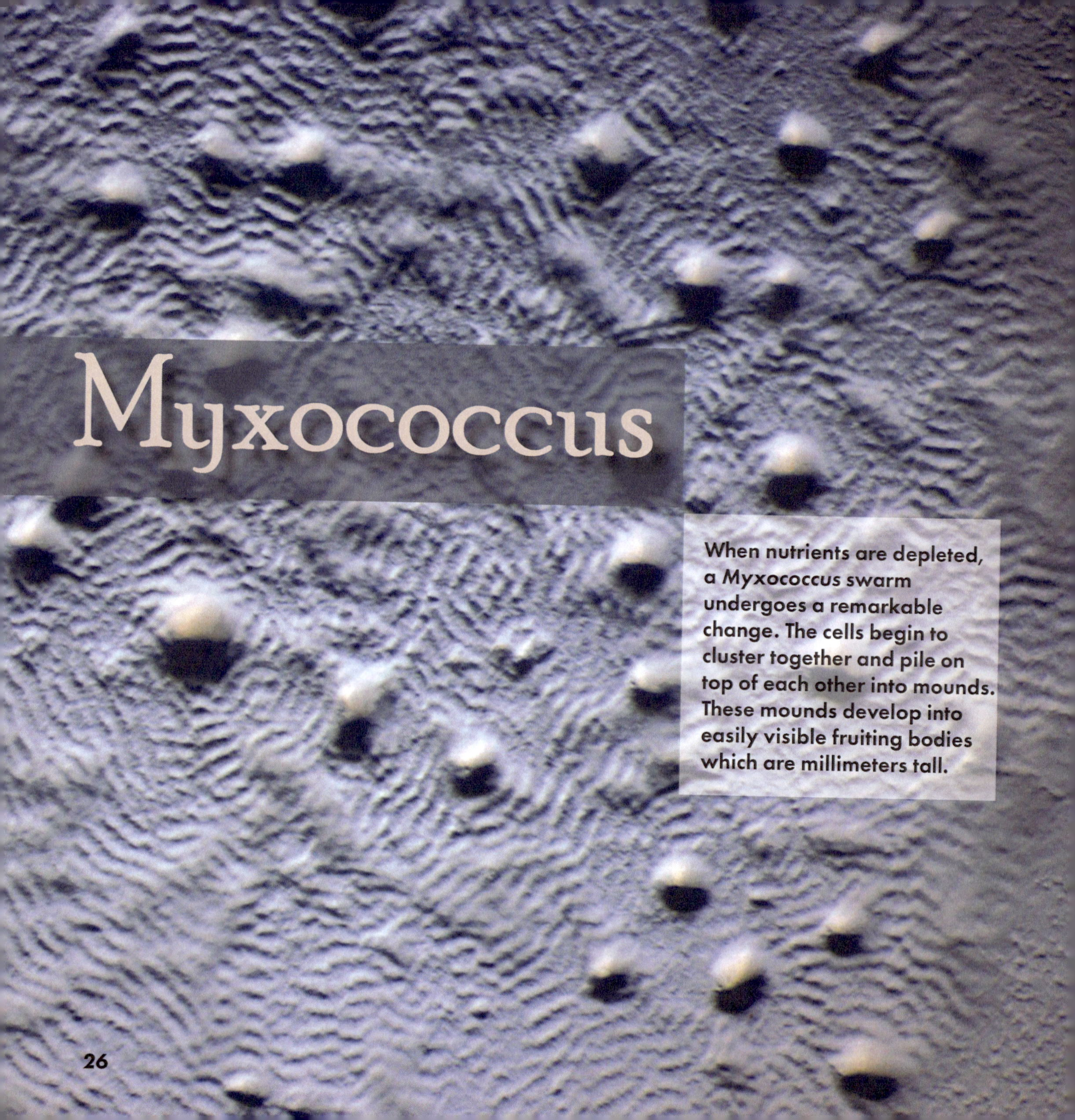

Myxococcus

When nutrients are depleted, a *Myxococcus* swarm undergoes a remarkable change. The cells begin to cluster together and pile on top of each other into mounds. These mounds develop into easily visible fruiting bodies which are millimeters tall.

Myxococcus commonly inhabits soil, where it primarily feeds on other microbial cells. When live prey is present, *Myxococcus* secretes enzymes which kill and break open these cells, releasing nutrients into the surrounding environment. When food is abundant, these bacteria move in swarms of cells, hunting and consuming prey. Each bacterium within the swarm glides on secreted slime in coordination with its neighbors.

Myxococcus fruiting bodies are constructed entirely of genetically identical *Myxococcus* cells, most of which die to provide a chance for some of the swarm to survive. The main structure of the fruiting body is constructed of dead cells. Just a small percentageof the bacteria become hardy myxospores which survive the lack of nutrients and germinate when nutrients become available again.

Myxococcus swarming and fruiting body formation depend on extensive coordination and cooperation. This is only possible with genetically identical cells, however. *Myxococcus* will not cooperate with genetically distinct individuals, even of the same species. If *Myxococcus* cells from two different genetic lineages are mixed, each fruiting body contains cells from only a single lineage.

M m

Cooperative Hunter

Without *Nitrosomonas* and *Nitrobacter*, the environment would be uninhabitable. They are integral to the nitrogen cycle, transformations of nitrogen necessary for life on the planet.

Nitrogen-containing waste is always entering the environment through the breakdown of organic materials, particularly protein, and the waste products of animals. A large fraction of the nitrogen waste is in the form of ammonia. As the concentration of ammonia increases, it can become toxic, especially in aquatic environments. *Nitrosomonas* and *Nitrobacter* work together to convert ammonia into a less dangerous form of nitrogen, nitrate. In addition to being less toxic than ammonia, nitrate is used by plants and most microbes as a nitrogen source.

Like plants, both *Nitrosomonas* and *Nitrobacter* build their cells from carbon dioxide rather than molecules obtained from other organisms. While plants get their building energy from the sun, *Nitrosomonas* and *Nitrobacter* use the energy they harvest from nitrogen compounds. They generate energy for their metabolism by extracting the electrons from ammonia or nitrite and transferring them to oxygen, similar to the way humans generate energy from the food we eat. In this process, *Nitrosomonas* uses ammonia and produces nitrite as waste. *Nitrobacter* then converts nitrite into nitrate.

Nitrobacter Nitrosomonas

nitrogen cyclers

Anyone who has ever had an aquarium is probably familiar with the importance of maintaining water quality. Ammonia is one of the biggest challenges. Ensuring that *Nitrosomonas* and *Nitrobacter* are present is critical to ammonia removal and a healthy aquarium.

Oenococcus

Producer of fine wines

Most of the fermentation involved in wine production is performed by yeast, but many wines, particularly red wines, benefit from additional fermentation by Oenococcus.

After the yeast fermentation has completed, *Oenococcus* can transform one of the common side products, malate, into lactic acid. This process, called malolactic fermentation, reduces the acidity of the wine, increases the flavor complexity, and improves the wine's resistance to microbial contamination and spoilage.

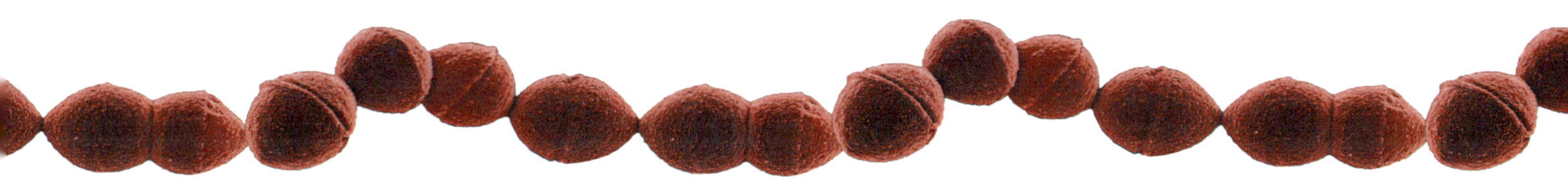

Like the yeast traditionally used in wine making, *Oenococcus* is naturally found on the skins of the grapes. *Oenococcus* is well-adapted to living in wine and thrives in the high alcohol, low pH environment. Naturally occurring *Oenococcus* can be used in the winemaking process, but commercially available cultures are usually used to maintain control over the process. If *Oenococcus* is not prpperly removed after fermentation, it can ruin the wine's taste and appearance.

P p

Prochloroccus is both the smallest and most abundant photosynthesizer on Earth. It is only 0.5 – 0.8 μm (millionths of a meter) in diameter, but the total mass of *Prochlorococcus* is estimated to equal about 900 Empire State Buildings. *Prochloroccus* covers the ocean surface around the world, with up to half a million in one teaspoon of seawater.

Prochlorococcus

In every breath you take...

Many people think of the tropical rain forests as being critically important to producing the oxygen found in the Earth's atmosphere. Trees do, in fact, contribute one quarter to one third of the Earth's oxygen. The ocean's phytoplankton, however, contributes half. *Prochlorococcus* alone produces one fifth of the oxygen you breathe, almost as much as all the trees on the planet. So, for one out of every five breaths you take, thank the humble *Prochlorococcus*.

Prochloroccus is a cyanobacterium, sometimes referred to as blue-green algae. Cyanobacteria are not technically algae, rather they are bacteria that, like plants, produce oxygen when they photosynthesize. Many bacteria can photosynthesize, but only cyanobacteria release oxygen as a waste product in the process. Photosynthesis in cyanobacteria and plants is very similar. In fact, the cell organelles where photosynthesis occurs in plants, chloroplasts, evolved from cyanobacteria billions of years ago when they formed a partnership with larger eukaryotic cells through a process called endosymbiosis.

Understanding *Quinella* could help reduce the greenhouse gases released by agriculture. *Quinella* is relatively common and abundant in ruminant animals. Although it is found in a wide variety of animals, it is most studied in domestic sheep. Sheep with high abundance of *Quinella* emit significantly less methane than most other sheep.

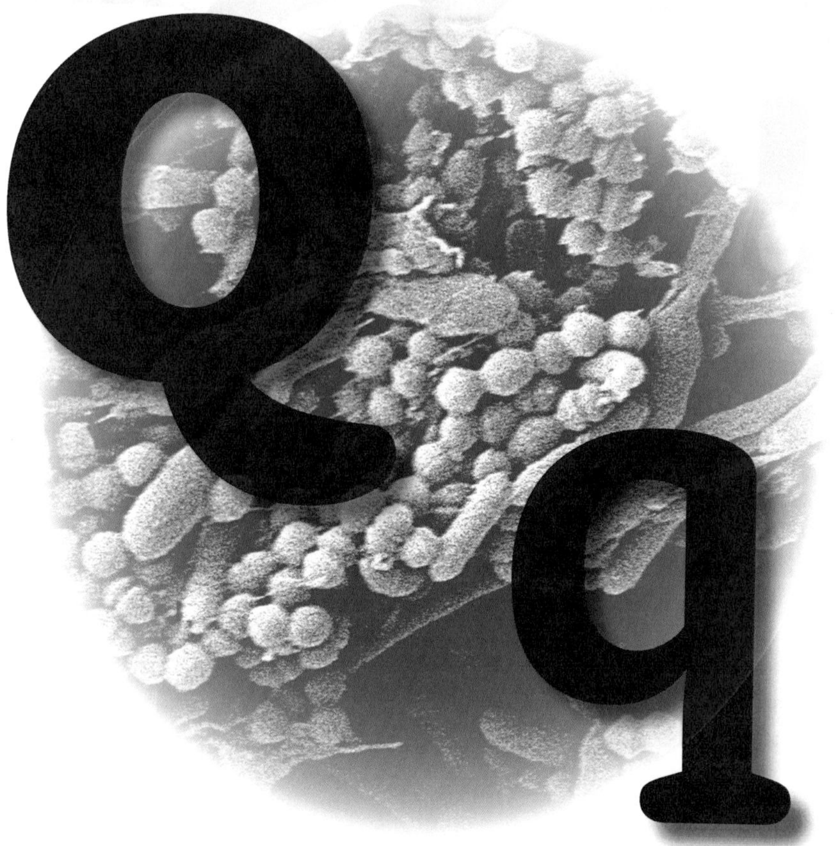

All animals have microbes in their digestive tracts that help them extract nutrients from food and regulate physiological processes. For animals that rely on high cellulose foods such as grass and hay, their resident microbes are indispensable. Animals cannot produce the enzymes necessary to break down cellulose. Many microbes adapted to living in the digestive tracts of cellulose-consuming animals, however, are masters of cellulose digestion. It is these microbes that digest the cellulose into sugars and other molecules which can enter and fuel the animal's cells.

Quinella

Rumen Resident

Ruminants are plant-eating animals like cows and sheep with a digestive organ called the rumen. Food swallowed by a ruminant first enters the rumen where a community of microbes breaks down cellulose from plants. Each microbe in the rumen has an important role in this digestive process. Some break apart the cellulose into sugars while others ferment those sugars. Some of the fermentation products are absorbed by the animal as food, but others, such as carbon dioxide and hydrogen, can build up and slow digestion. Critical rumen microbes called methanogens use the carbon dioxide and hydrogen and release methane. Methane production is necessary to keep digestion going in ruminants, but methane is also a potent greenhouse gas.

Although first observed in 1913, *Quinella* still cannot be cultured, precluding lab studies of how it might reduce methane emissions. Its genome sequence, however, reveals that when *Quinella* ferments it does not produce carbon dioxide or hydrogen, the fermentation waste products used by methanogens to produce methane.

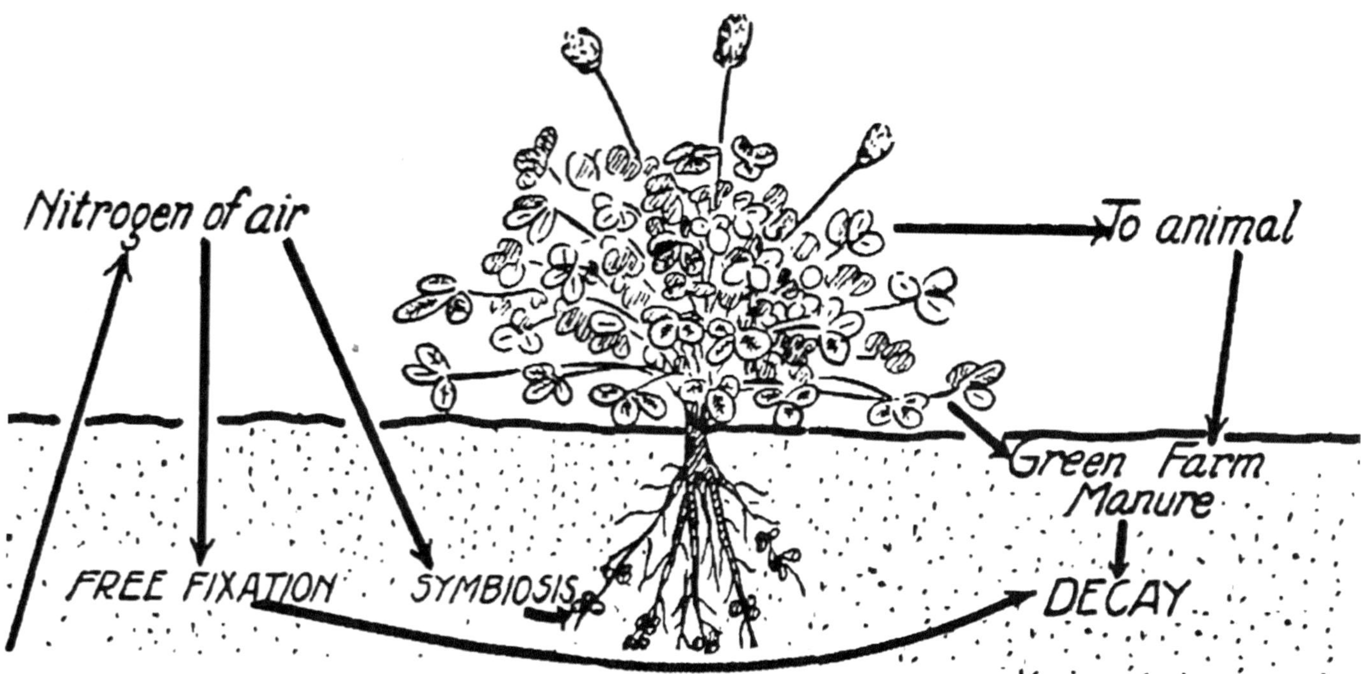

Rhizobium

Nitrogen is required to build proteins, DNA, and other molecules essential to all living organisms. Although 70% of Earth's atmosphere is nitrogen gas, it is so chemically stable that only a few bacteria and archaea have evolved the energy-intensive metabolism necessary to convert nitrogen gas into a biologically usable form. Prior to the invention of synthetic fertilizers, these select few nitrogen-fixing microbes were essentially the only way that the living world acquired nitrogen.

Nature's Fertilizer

Some nitrogen-fixers are free-living, but *Rhizobium* forms a symbiotic relationship with legumes such as peas, beans, and clover. When these plants are starved for nitrogen, *Rhizobium* in the soil is attracted to chemical signals that they release from their roots. The *Rhizobium* enters the plant root and forms a nodule. In the nodule, the plant protects the *Rhizobium* from oxygen (which interferes with nitrogen-fixation) by producing a protein called leghemoglobin. This protein is closely related to the hemoglobin which carries oxygen in your blood. With the plant's protection, *Rhizobium* can convert atmospheric nitrogen into a form the plant can use. In return, the plant provides nutrition for the *Rhizobium*.

The practice of crop rotation existed for thousands of years before anything was known about *Rhizobium* or nitrogen-fixation. By alternately planting nitrogen-hungry crops like corn with legumes in the same field, the nitrogen in the soil can be restored by the legumes' symbiotic nitrogen-fixing *Rhizobium*. This way farmers can maintain healthy soil and successfully grow a variety of crops without the need for synthetic fertilizer.

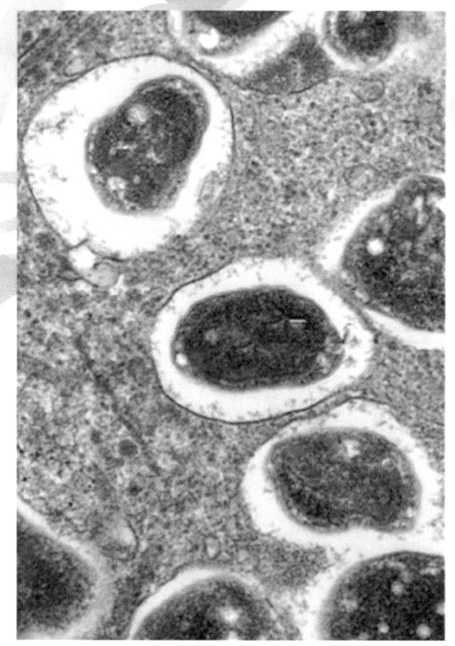

Rhizobium and its close relatives are perhaps the most agriculturally important nitrogen-fixing bacteria.

Streptomyces

Researchers often look to *Streptomyces* for drug discovery. Typically living in soil, leaf litter, and rotting logs, *Streptomyces* is in competition with many other bacteria and fungi for the same resources. As part of its growth cycle, *Streptomyces* produces many secondary metabolites, chemical compounds not necessary for survival but which can provide a competitive advantage. These compounds have a wide range of functions for the bacteria, from acquiring nutrients and protection from environmental stresses to killing or communicating with other microbes. Many have unknown functions, but almost all can alter or kill the cells of other organisms.

Antibiotics and Aromas

Streptomyces are the source of more than a dozen antibiotics and other important drugs. Some familiar examples include streptomycin, chloramphenicol, and neomycin.

One specific compound from *Streptomyces* called geosmin is a main ingredient in the distinct smell of soil and rain. In fact, most things with an earthy taste or smell contain geosmin. It is even added to perfumes. Next time you take a walk in the woods or go outside after a spring rain, take a deep breath and appreciate the aromas brought to you by *Streptomyces*.

In culture, *Streptomyces* is one of the most beautiful types of bacteria. Its fuzzy, spore-containing colonies come in almost every color of the rainbow, including the sky-blue *S. coelicolor* (pictured here), bright pink *S. aurantiacus*, and orange *S. spectabilis*. In addition to being beautiful, the pigments produced by *Streptomyces* protect it from the damaging effects of UV light.

Thermus (as well as *Deinococcus* (p. 8)) are among the oldest groups of bacteria. Studying their biology can help us understand the earliest life on Earth.

Thermus
The Spring of Biotechnology

The process of polymerase chain reaction (PCR) was invented in the early 1980s. This technique is used to produce billions of copies of one specific region of DNA. It is critical to biotechnology research and has many applications in medicine (e.g. COVID testing), food safety (e.g. detecting contamination), and forensic investigations (e.g. DNA fingerprinting).

PCR requires the temperature of the reaction to be repeatedly raised to near boiling. The enzyme which builds the new DNA molecules (DNA polymerase) from most organisms is destroyed by such high temperatures. Enzymes are proteins, and above a certain temperature any protein will lose its shape and no longer work. The natural habitat of *Thermus aquaticus*, however, is hot springs with temperatures approaching boiling. Its DNA polymerase, typically referred to as *taq* polymerase, can therefore keep working at the high temperatures necessary for PCR. Although PCR has been refined and improved over the last 40 years, *taq* polymerase is still the most commonly used enzyme for the process.

Anywhere on Earth that there is water and a way to obtain nutrients and energy, microbes have been found.

Even 10 km above the surface of the Earth there are about 5000 bacteria per cubic meter of air. These bacteria are carried by wind currents through the atmosphere on dust particles, water droplets, and as single cells. Some atmospheric bacteria are just there by chance, picked up by winds from the land or sea, but others are adapted to grow in the upper atmosphere.

Bacteria and other microbes have a large effect on weather patterns. More than half the raindrops and snowflakes that fall are formed around bacteria, and the formation of clouds themselves rely in part on airborne bacteria.

Ubiquitous

On the surface of deserts, bacteria form crusts which prevent evaporation, stabilize soil, and provide nutrients making it possible for plants to gain a foothold in these harsh environments.

More than one and a half miles down below the Earth's surface, bacteria have been found living in isolated pockets of water in rock.

In the deepest depths of the ocean, life thrives around geothermal vents where bacteria use chemical energy instead of light to form the base of the food chain.

Vibrio

Visible from space

For centuries, sailors told tales of milky seas where the ocean would glow with a blue-white light from horizon to horizon. Herman Melville, Jules Verne, and Charles Darwin all remark on the phenomenon in their writing. It wasn't until 2005, however, that the bioluminescent bacterium *Vibrio* was identified as the cause.

In 1995, a report of milky seas off the coast of Somalia was matched with satellite images of a glowing patch of ocean the size of Connecticut. Further investigation estimated that 40 billion million million (4×10^{22}) *Vibrio* cells simultaneously bioluminescing were responsible. The cause of the immense bacterial growth at the root of the milky seas is still unknown.

Light production by *Vibrio* was the first bacterial behavior determined to be controlled by a system called quorum sensing. In quorum sensing, chemical signals are produced and sensed by bacteria to measure thier population size and alter behavior accordingly. We now know that quorum sensing regulates many different bacterial behaviors including causing disease, producing antibiotics (*e.g. Streptomyces* (p. 38)), and changing life cycle stages (e.g. endospore production by *Clostridium* (p.6) and fruiting body formation by *Myxococcus* (p.26)).

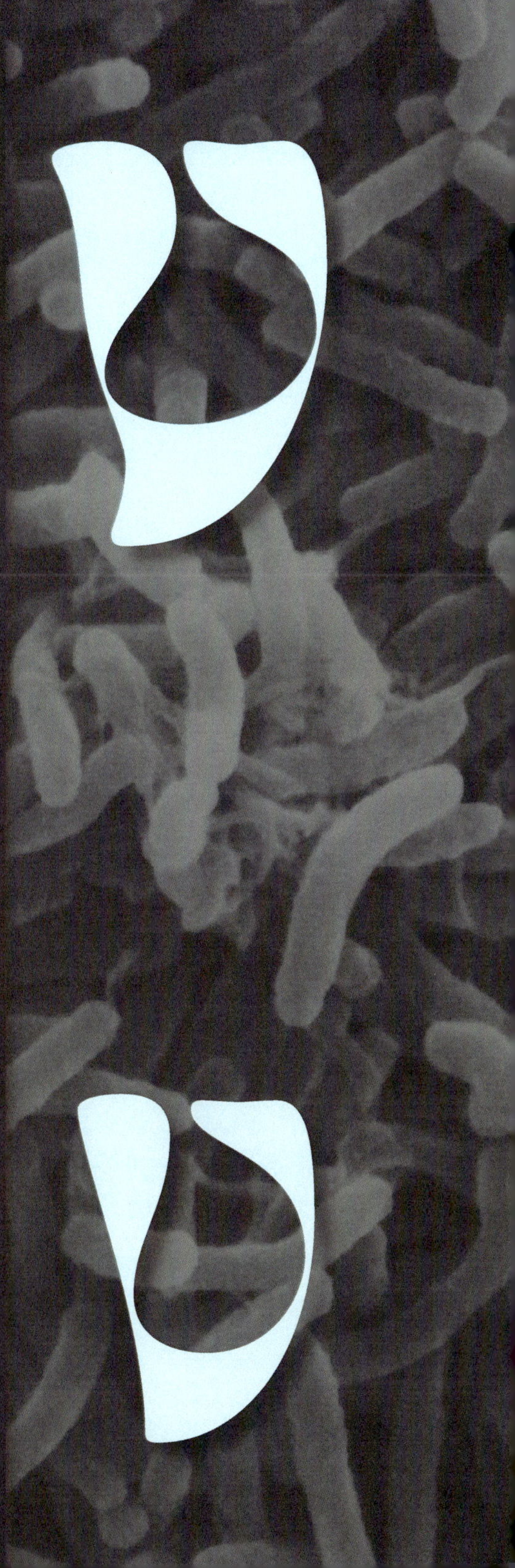

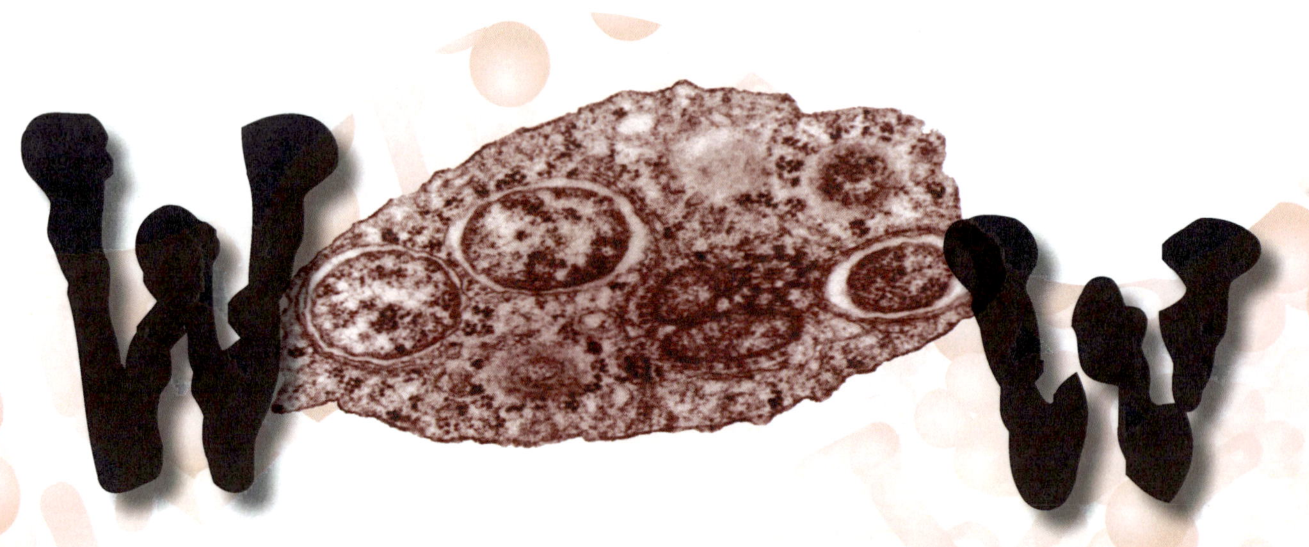

At least half of all insect species harbor the bacterium *Wolbachia* inside their cells. These bacteria cannot live outside an insect host and are passed directly from mother to offspring in her eggs. By controlling male reproduction and survival, *Wolbachia* maintains its own prevalence within a population.

Wolbachia controls insect reproduction for its own benefit in several ways. Because males cannot directly pass the bacterium to the next generation infected males are sometimes feminized or simply killed. In some cases, infected males can only successfully mate with infected females. Certain species of insects have become almost entirely female and reproduce by parthenogenesis, a type of reproduction where the female produces offspring without fertilization by a male.

One promising application of *Wolbachia* is mosquito control. By releasing *Wolbachia*-infected males, fewer females will successfully reproduce because they do not naturally harbor *Wolbachia*. Also, if *Wolbachia* does become established in a mosquito population, there is still a benefit. Mosquitoes infected with *Wolbachia* are less likely to spread viral diseases such as Zika, dengue, chikungunya, and yellow fever.

Xanthomonas species cause over 350 different plant diseases, affecting some of the world's most important crops.

Rice, citrus, sugar cane, bananas, and a wide variety of vegetables can all fall victim to *Xanthomonas*. These diseases can have a huge economic impact. For example, banana wilt alone, if left unchecked, could cause over $25 billion in damage.

Xanthomonas

Xanthomonas can cause disease either by infecting the photosynthetic cells of the plant's leaves (such as citrus canker) or blocking the plant's water transport system, the xylem (such as black rot in cabbages and related vegetables). As the disease progresses, the bacteria reach the surface of the plant and are spread by wind and rain. Often *Xanthomonas* is introduced through contaminated seeds, making proper treatment and importation of seeds critical to preventing crop damage by these diseases.

You have probably eaten the xylem-blocking substance produced by *Xanthomonas*. This water-soluble gel, xanthan gum, is produced industrially by *Xanthomonas campestris*. It is used as a thickening agent in wide variety of foods and can provide texture in gluten-free products.

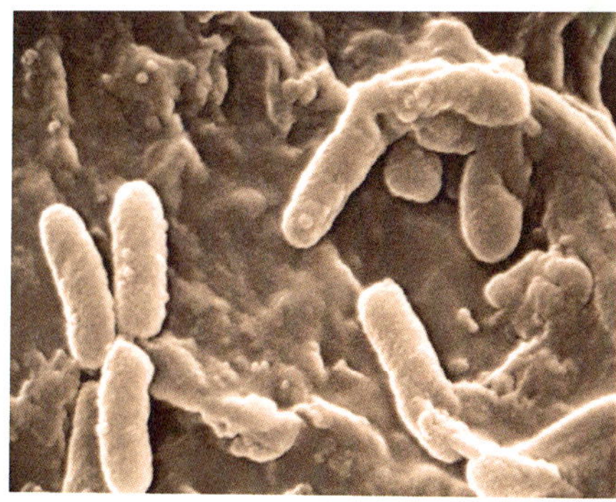

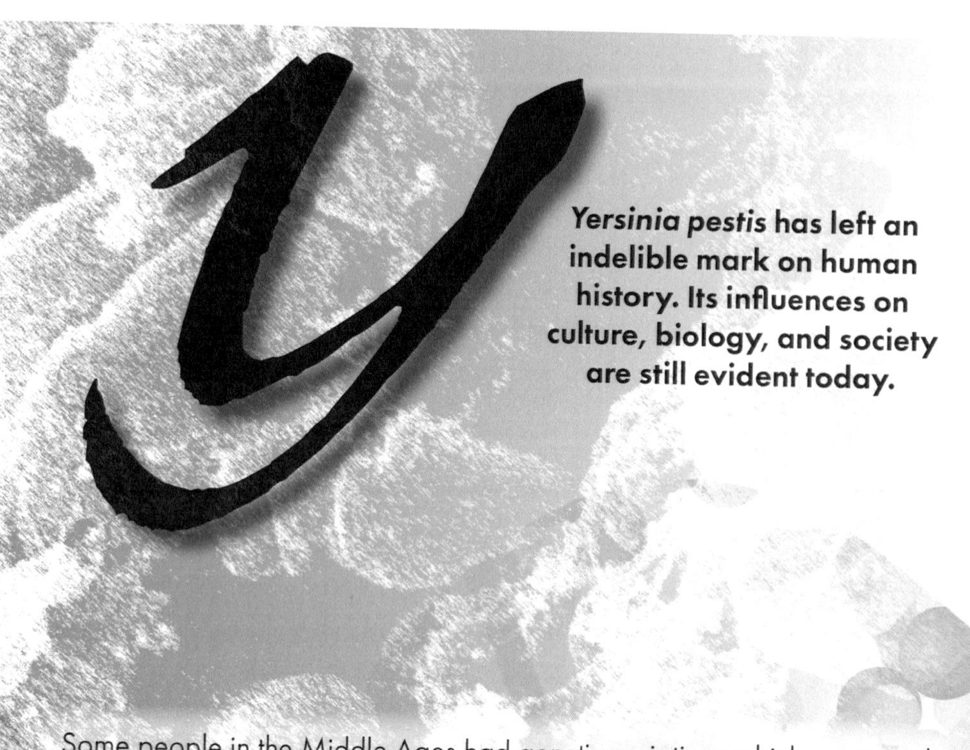

Yersinia pestis has left an indelible mark on human history. Its influences on culture, biology, and society are still evident today.

Because so many workers died from plague the number of available workers drastically decreased. Workers began to demand higher wages and aspire for greater comfort and respect. Eventually the feudal system which dominated Europe for hundreds of years collapsed.

Some people in the Middle Ages had genetic variations which prevented *Yersinia pestis* from entering lymph nodes or allowed the immune system to attack the bacterium more quickly. These people were more likely to survive and passed on their genes. As a result, these protective genetic variations became more common in Europeans following the Black Death. Today, these genetic variations can help protect against other infections, including HIV and COVID-19.

If you ever say "God bless you" when someone sneezes or play "Ring Around the Rosie" you are continuing traditions begun because of the plague.

Yersinia
Bringer of the Black Death

Yersinia pestis is the bacterium that causes plague. Also known as the Black Death, this disease devastated Europe in the Middle Ages. Arriving from Asia in Genoa in 1347, plague spread throughout Europe over the next decades ultimately killing one third of the population.

Bubonic plague occurs when *Yersinia pestis* enters the body through direct contact or a flea bite. Once in the body, the bacteria travel to the lymph nodes. There they multiply and cause extraordinary swelling of the lymph nodes, called "buboes". More than half of untreated bubonic plague victims die within a week. If inhaled, *Y. pestis* results in pneumonic plague which causes flu-like symptoms, pneumonia, and, without treatment, almost always death. If the bacteria reach the bloodstream, blood supply becomes restricted resulting in gangrene and, like pneumonic plague, is almost always fatal in untreated patients.

The future of biofuel

As society moves away from fossil fuels, the development of alternatives is necessary. One possible alternative fuel is ethanol. *Zymomonas* can convert sugars to ethanol with almost 100% efficiency, far better than any other studied microbe.

Sugars usually come from sources that are also used for food. Ideally a starting material other than sugars would be used to make ethanol, such as the parts of crops typically treated as waste. Much of this plant waste is cellulose, the tough material that makes up plant cell walls. Even though cellulose is constructed from sugars, it is difficult to break down.

Through genetic engineering *Zymomonas* can be given the ability to produce ethanol from cellulose, although not as well as from sugars. Many researchers are working to improve this process.

In addition to fuel, oil and natural gas are used to make many chemicals used in industrial applications. One example is 2,3-butanediol, used to manufacture a wide variety of products including inks, synthetic rubber, cosmetics, and food additives. Similar to ethanol production, *Zymomonas* is being explored as an alternative source of 2,3-butanediol and potentially other industrially important chemicals.

Image Sources

The images in this book have been modified from those listed here. Credits are in order of appearance. Rarely images of the subject bacteria were not available. In those cases, images of visually similar bacteria were used.

Vader1941. 2016. M luteus visualized using scanning electron microscopy. CC BY-SA 4.0 DEED. https://en.wikipedia.org/wiki/File:M_luteus_visualized_using_scanning_electron_microscopy.png.

Das Murtey, M and P Ramasamy. 2016. Lactobacillus acidophilus SEM. CC BY-SA 3.0 DEED. https://commons.wikimedia.org/wiki/File:Lactobacillus_acidophilus_SEM.jpg.

Matthysse, AG, K V Holmes, and RHG Gurlitz. 2008. Agrobacterium tumefaciens. Public Domain. https://commons.wikimedia.org/w/index.php?curid=34654351.

Anonymous. Date unknown. Tree face fairy tale. Free to Use. https://www.pikist.com/free-photo-itycq.

NIAID. 2012. E coli. CC BY-SA 2.0 DEED. https://www.flickr.com/photos/niaid/7316101966.

Elionas2. Date unknown. instetines with person. Free to Use. https://pixabay.com/illustrations/offal-marking-medical-colon-liver-1463369/.

CDC. 2016. clostridium cdc. Public Domain. https://picryl.com/media/clostridium-cdc-21911-77c8cd.

Anonymous. Date unknown. Compost. Free to Use. https://www.pickpik.com/green-waste-compost-compost-bin-bunch-fruit-peels-132514.

Catalania. Date unknown. radiation symbol danger. Free to Use. https://pixabay.com/illustrations/radiation-symbol-danger-646217/.

Daly, M. 2008. Deinococcus. Public Domain. https://www.wpafb.af.mil/News/Photos/igphoto/2000655676/.

Angert, E. 2002?. epulopscium-1. By permission. https://cals.cornell.edu/microbiology/research/active-research-labs/angert-lab/epulopiscium.

Zerpe, R. 2019. Surgeonfish. CC BY 2.0 DEED. https://commons.wikimedia.org/wiki/File:Yellowfin_surgeonfish_%28Acanthurus_xanthopterus%29_%2848564924662%29.jpg.

Kauerhoff, T. 2017. Curious Turtle A Curtle. CC BY 3.0 DEED. https://commons.wikimedia.org/wiki/File:Curious_Turtle_A_Curtle_%28225008535%29.jpeg.

NIAID. 2015. Macrophage Infected with Francisella tularensis Bacteria. CC BY 2.0 DEED. https://commons.wikimedia.org/wiki/File:Macrophage_Infected_with_Francisella_tularensis_Bacteria_%285950310835%29.jpg.

Harrison, JJ. 2009. Oryctolagus cuniculus Tasmania. CC BY-SA 3.0 DEED. https://upload.wikimedia.org/wikipedia/commons/3/37/https://commons.wikimedia.org/wiki/File:Oryctolagus_cuniculus_Tasmania_2.jpg.

Tann, J. 2011. Tick. CC BY 2.0 DEED. https://commons.wikimedia.org/wiki/File:Tick_%286368335425%29.jpg.

CNX OpenStax. 2015. Geobacter. CC BY 4.0 DEED. https://commons.m.wikimedia.org/wiki/File:OSC_Microbio_04_00_splash.jpg.

Cossiedog. 2013. Diagram of electron flow in a biological photovoltaic system. CC BY-SA 3.0 DEED. https://commons.wikimedia.org/wiki/File:Diagram_of_electron_flow_in_a_biological_photovoltaic_system.jpg.

Rosatti, L. Date unknown. uyuni salt flat. Free to Use. https://www.pexels.com/photo/uyuni-salt-flat-2613110/.

Rotational~commonswiki. 2008. Haloquadratum walsbyi. https://commons.wikimedia.org/wiki/File:Haloquadratum_walsbyi00_%28cropped%29.png.

Pesterev, S. 2017. dune and sky. Public Domain. https://unsplash.com/photos/desert-during-nighttime-XaidrBZfEwU.

Xu, M. 2020. Janthinobacterium aquaticum. https://commons.wikimedia.org/wiki/File:Janthinobacterium_aquaticum.jpg.

Mallette, L. Date unknown. purple abstract background. Public Domain. https://www.publicdomainpictures.net/en/view-image.php?image=266016&picture=purple-abstract-background.

Anonymous. Date unknown. palm skin. Free to Use. https://www.wallpaperflare.com/id-human-body-part-close-up-full-frame-human-skin-one-person-wallpaper-ailuc.

Daderot. 2018. Charcuterie with berries. Public Domain. https://commons.wikimedia.org/wiki/File:Charcuterie_with_berries_-_Cliff_House_Maine_-_Cape_Neddick,_Maine_-_20180723_141828.jpg.

Gemini, T. 2008. Myxococcus xanthus. CC BY-SA 3.0 DEED. https://commons.wikimedia.org/wiki/File:Myxococcus_xanthus_%281%29.png.

KoiQuestion. 2012. nitrosomonas. CC BY-SA 2.0 DEED. https://www.flickr.com/photos/koiquest10/6876052456/in/photostream/.

Janine. 2011. Fish tank. CC BY 2.0 DEED. https://commons.wikimedia.org/wiki/File:Fish_tank_%282%29.jpg.

NIAID. 2023. S pyogenes SEM. CC BY 2.0 DEED. https://www.flickr.com/photos/niaid/52602981880/in/photostream/.

Yashima. 2006. Pour me some wine please. CC BY-SA 2.0 DEED. https://commons.wikimedia.org/wiki/File:Pour_me_some_wine_please_-_Flickr_-_yashima.jpg.

Kratochvil, P. Date unknown. ocean surface. Public Domain. https://www.publicdomainpictures.net/en/view-image.php?image=4775&picture=sea-surface.

Thompson, L and N Watson. 2007. Prochlorococcus marinus. Public Domain. https://commons.wikimedia.org/wiki/File:Prochlorococcus_marinus.jpg.

Zacharski, KA. 2017. Mixed culture biofilm. CC BY 4.0 DEED. https://commons.wikimedia.org/wiki/File:Mixed-culture_biofilm.jpg.

Drichards2. 2008. Cotswold Sheep. CC BY-SA 3.0 DEED. https://upload.wikimedia.org/wikipedia/commons/2/29/Cotswold_Sheep_%28cropped%29.JPG.

CNX OpenStax. 2016. Rhizobium symbiosis. Public Domain. https://commons.wikimedia.org/wiki/File:Figure_31_03_02ab.jpg.

Lyon, TL. 1918. nitrogen cycle diagram. Public Domain. https://www.flickr.com/photos/internetarchivebookimages/14594696440.

Docwarhol. 2017. Streptomyces griseus color enhanced scanning electron micrograph. CC BY-SA 4.0 DEED. https://commons.wikimedia.org/wiki/File:Streptomyces_griseus_color_enhanced_scanning_electron_micrograph..jpg.

Yana-t Petruk, T. 2015. Колонії Streptomyces. CC BY-SA 4.0 DEED. https://commons.wikimedia.org/

Salvagnin, D. Date unknown. Thermus Black Sand Basin. CC BY 2.5 DEED. https://www.bibalex.org/SCIplanet/en/Article/Details?id=13761.

Anonymous. 2017. Therm. CC BY-SA 4.0 DEED. https://commons.wikimedia.org/wiki/File:Therm.png.

NASA. 2016. California coastal current. Public Domain. https://earthobservatory.nasa.gov/images/87575/california-coastal-current.

Miller, S. 1995. Milky sea nightcolor. Public Domain. https://en.wikipedia.org/wiki/Milky_seas_effect#/media/File:Milky_sea_nightcolor5x5.jpg.

Taylor, R, T Kirn, and L Howard. 2005. Cholera bacteria SEM. Public Domain. https://commons.wikimedia.org/wiki/File:Cholera_bacteria_SEM.jpg.

O'Neill, S. 2004. Wolbachia. CC BY 2.5 DEED. https://en.m.wikipedia.org/wiki/File:Wolbachia.png.

Defense Visual Information Distribution Service. 2006. a female aedes aegypti mosquito. Public Domain. https://nara.getarchive.net/amp/media/a-female-aedes-aegypti-mosquito-while-she-was-in-the-d28428.

Nelson, S. 2018. xanthomonas infection. Public Domain. https://www.flickr.com/photos/scotnelson/41535429755/in/photostream/.

CDC/ Janice Haney Carr. 2006. Pseudomonas aeruginosa SEM. Public Domain. https://commons.wikimedia.org/wiki/File:Pseudomonas_aeruginosa_SEM.jpg.

Anonymous. unknown. Yersinia pestis. Public Domain. https://pixy.org/5872931/.

darksouls1 (original engraving Furst, P. 17th Century). unknown. mask. Free to Use. https://pixabay.com/illustrations/mask-doctor-plague-doctor-plague-5770221/.

Anonymous. unknown. automobile vehicle transport gasoline. Free to Use. https://www.pxfuel.com/en/free-photo-opmvl.

Eder, W. 2011. Undibacterium oligocarboniphilum. CC BY 4.0 DEED. https://commons.wikimedia.org/wiki/File:Undibacterium_oligocarboniphilum.jpg.

References

An S-Q, Potnis N, Dow M, Vorhölter F-J, He Y-Q, Becker A, Teper D, Li Y, Wang N, Bleris L, *et al.* 2019. Mechanistic insights into host adaptation, virulence and epidemiology of the phytopathogen *Xanthomonas*. FEMS Microbiol Rev. 44(1):1–32.

Anonymous. 2015 Jan 6. Brassicas, Black Rot. Center for Agriculture, Food, and the Environment. https://ag.umass.edu/vegetable/fact-sheets/brassicas-black-rot.

Baricz A, Teban A, Chiriac CM, Szekeres E, Farkas A, Nica M, Dascălu A, Oprișan C, Lavin P, Coman C. 2018. Investigating the potential use of an Antarctic variant of *Janthinobacterium* lividum for tackling antimicrobial resistance in a One Health approach. Sci Rep. 8(1):15272.

Bech-Terkilsen S, Westman JO, Swiegers J, Siegumfeldt H. 2020. *Oenococcus oeni*, a species born and moulded in wine: a critical review of the stress impacts of wine and the physiological responses. Australian Journal of Grape and Wine Research. 26(3):188–206.

Beyond Our Solar System. NASA Solar System Exploration. https://solarsystem.nasa.gov/solar-system/beyond/overview.

Biakowska AM. 2016. Strategies for efficient and economical 2,3-butanediol production: new trends in this field. World Journal of Microbiology and Biotechnology. 32(12):1–14.

Braga A, Gomes D, Rainha J, Amorim C, Cardoso BB, Gudiña EJ, Silvério SC, Rodrigues JL, Rodrigues LR. 2021. *Zymomonas mobilis* as an emerging biotechnological chassis for the production of industrially relevant compounds. Bioresources and Bioprocessing. 8(1):128.

CDC. 2022 Jul 25. Mosquitoes with *Wolbachia* | CDC. Centers for Disease Control and Prevention. https://www.cdc.gov/mosquitoes/mosquito-control/community/emerging-methods/wolbachia.html.

CDC. 2022 Jul 6. National Infection & Death Estimates for AR. Centers for Disease Control and Prevention. https://www.cdc.gov/drugresistance/national-estimates.html.

CDC Tularemia | FAQ About Tularemia. 2019 Feb 21. https://emergency.cdc.gov/agent/tularemia/faq.asp.

Chen YE, Fischbach MA, Belkaid Y. 2018. Skin microbiota–host interactions. Nature. 553(7689):427–436.

Chivian D, Brodie EL, Alm EJ, Culley DE, Dehal PS, DeSantis TZ, Gihring TM, Lapidus A, Lin L-H, Lowry SR, et al. 2008. Environmental genomics reveals a single-species ecosystem deep within Earth. Science. 322(5899):275–278.

Christner BC, Cai R, Morris CE, McCarter KS, Foreman CM, Skidmore ML, Montross SN, Sands DC. 2008. Geographic, seasonal, and precipitation chemistry influence on the abundance and activity of biological ice nucleators in rain and snow. Proc Natl Acad Sci USA. 105(48):18854–18859.

Davis CP. 1996. Normal Flora. In: Baron S, editor. Medical Microbiology. 4th ed. Galveston (TX): University of Texas Medical Branch at Galveston. http://www.ncbi.nlm.nih.gov/books/NBK7617/.

Doucleff M. 2022 Oct 21. Black Death survivors gave their descendants a genetic advantage — but with a cost. NPR. ://www.npr.org/sections/goatsandsoda/2022/10/19/1129965424/how-black-death-survivors-gave-their-descendants-an-edge-during-pandemics.

Dykhuizen D. 2005. Species numbers in bacteria. Proc Calif Acad Sci. 56(6 Suppl 1):62–71.

Glatter KA, Finkelman P. 2021. History of the plague: An ancient pandemic for the age of COVID-19. Am J Med. 134(2):176–181.

Grice EA, Kong HH, Renaud G, Young AC, Bouffard GG, Blakesley RW, Wolfsberg TG, Turner ML, Segre JA. 2008. A diversity profile of the human skin microbiota. Genome Res. 18(7):1043–1050.

Hampton-Marcell JT, Larsen P, Anton T, Cralle L, Sangwan N, Lax S, Gottel N, Salas-Garcia M, Young C, Duncan G, et al. 2020. Detecting personal microbiota signatures at artificial crime scenes. Forensic Science International. 313:110351.

Holkar S, Begde D, Nashikkar N, Kadam T, Upadhyay A. 2013. Optimization of some culture conditions for improved biomass and antibiotic production by streptomyces spectabilis isolated from soil. https://www.semanticscholar.org/paper/optimization-of-some-culture-conditions-for-biomass-Holkar-Begde/450d7afd024fdabdeec69cc6695e5883b7466d92.

How much oxygen comes from the ocean? ://oceanservice.noaa.gov/facts/ocean-oxygen.html.

Jones RM, Nicas M, Hubbard A, Sylvester MD, Reingold A. 2005. The infectious dose of *Francisella tularensis* (Tularemia). Appl Biosaf. 10(4):227–239.

Kanelli M, Mandic M, Kalakona M, Vasilakos S, Kekos D, Nikodinovic-Runic J, Topakas E. 2018. Microbial production of violacein and process optimization for dyeing polyamide fabrics with acquired antimicrobial properties. Frontiers in Microbiology. 9. https://www.frontiersin.org/articles/10.3389/fmicb.2018.01495.

King CH, Desai H, Sylvetsky AC, LoTempio J, Ayanyan S, Carrie J, Crandall KA, Fochtman BC, Gasparyan L, Gulzar N, et al. 2019. Baseline human gut microbiota profile in healthy people and standard reporting template. PLOS ONE. 14(9):e0206484.

Kong HH, Segre JA. 2012. Skin microbiome: Looking back to move forward. Journal of Investigative Dermatology. 132(3):933–939.

Kooken JM, Fox KF, Fox A. 2012. Characterization of *Micrococcus* strains isolated from indoor air. Molecular and Cellular Probes. 26(1):1–5.

Krulwich R. 2012 Sep 17. Which is greater, the number of sand grains on earth or stars in the sky? NPRhttps://www.npr.org/sections/krulwich/2012/09/17/161096233/which-is-greater-the-number-of-sand-grains-on-earth-or-stars-in-the-sky.

Kumar S, Altermann E, Leahy SC, Jauregui R, Jonker A, Henderson G, Kittelmann S, Attwood GT, Kamke J, Waters SM, et al. 2022. Genomic insights into the physiology of *Quinella*, an iconic uncultured rumen bacterium. Nat Commun. 13(1):6240.

Levin PA, Angert ER. 2015. Small but Mighty: Cell Size and Bacteria. Cold Spring Harb Perspect Biol. 7(7):a019216.

Lorentzen MPG, Lucas PM. 2019. Distribution of *Oenococcus oeni* populations in natural habitats. Appl Microbiol Biotechnol. 103(7):2937–2945.

Makarova KS, Aravind L, Wolf YI, Tatusov RL, Minton KW, Koonin EV, Daly MJ. 2001. Genome of the extremely radiation-resistant bacterium *Deinococcus radiodurans* viewed from the perspective of comparative genomics. Microbiol Mol Biol Rev. 65(1):44–79.

Marco ML, Heeney D, Binda S, Cifelli CJ, Cotter PD, Foligné B, Gänzle M, Kort R, Pasin G, Pihlanto A, et al. 2017. Health benefits of fermented foods: microbiota and beyond. Current Opinion in Biotechnology. 44:94–102.

Marin VR, Ferrarezi JH, Vieira G, Sass DC. 2019. Recent advances in the biocontrol of *Xanthomonas spp.* World J Microbiol Biotechnol. 35(5):72.

Meet the Molecules; Geosmin. 2017 Nov 1. John Innes Centre. https://www.jic.ac.uk/blog/meet-the-molecules-geosmin/.

Microbes Survive, and Maybe Thrive, High in the Atmosphere. https://www.science.org/content/article/microbes-survive-and-maybe-thrive-high-atmosphere.

Miller SD, Haddock SHD, Elvidge CD, Lee TF. 2005. Detection of a bioluminescent milky sea from space. Proceedings of the National Academy of Sciences. 102(40):14181–14184.

Muñoz-Dorado J, Marcos-Torres FJ, García-Bravo E, Moraleda-Muñoz A, Pérez J. 2016. Myxobacteria: Moving, killing, feeding, and surviving together. Front Microbiol. 7:781.

Oubre CM. Microbial Monitoring of the International Space Station.

Partensky F, Hess WR, Vaulot D. 1999. *Prochlorococcus*, a marine photosynthetic prokaryote of global significance. Microbiol Mol Biol Rev. 63(1):106–127.

Petsakos A, Kozicka M, Blomme G, Nakakawa JN, Ocimati W, Gotor E. 2023. The potential impact of banana *Xanthomonas* wilt on food systems in Africa: modeling scenarios of policy response and disease control measures. Frontiers in Sustainable Food Systems. 7. https://www.frontiersin.org/articles/10.3389/fsufs.2023.1207913.

Pierson DL, Botkin DJ, Bruce RJ, Castro VA, Smith MJ, Oubre CM, Ott CM. 2013. Microbial monitoring of the International Space Station. Osaka. https://ntrs.nasa.gov/citations/20130013534.

Prokaryotes: The unseen majority. https://www.pnas.org/doi/10.1073/pnas.95.12.6578.

Quinn GA, Banat AM, Abdelhameed AM, Banat IM. 2020. *Streptomyces* from traditional medicine: sources of new innovations in antibiotic discovery. J Med Microbiol. 69(8):1040–1048.

Roth RR, James WD. 1988. Microbial ecology of the skin. Annual Review of Microbiology. 42(1):441–464.

The Black Death. Historic UK. http://www.historic-uk.com/HistoryUK/HistoryofEngland/The-Black-Death/.

Todhanakasem T, Wu B, Simeon S. 2020. Perspectives and new directions for bioprocess optimization using *Zymomonas mobilis* in the ethanol production. World Journal of Microbiology and Biotechnology. 36(8).

Wang J, Ren K, Zhu Y, Huang J, Liu S. 2022. A review of recent advances in microbial fuel cells: Preparation, operation, and application. BioTech (Basel). 11(4):44.

What You Didn't Know About *Janthinobacterium*. Small Things Considered. https://schaechter.asmblog.org/schaechter/2009/04/by-jenna-tabor-godwin-rhona-stuart-rosa-i-le%C3%B3n-zayas-and-chitra-rajakuberan--------janthinobacterium-cultured-on-nb-aga.html.

Wolbachia. http://www.eliminatedengue.com/our-research/wolbachia.